AF228481

Ermine

by Grace Hansen

Abdo Kids Jumbo is an Imprint of Abdo Kids
abdobooks.com

abdobooks.com

Published by Abdo Kids, a division of ABDO, P.O. Box 398166, Minneapolis, Minnesota 55439.
Copyright © 2020 by Abdo Consulting Group, Inc. International copyrights reserved in all countries.
No part of this book may be reproduced in any form without written permission from the publisher.
Abdo Kids Jumbo™ is a trademark and logo of Abdo Kids.

Printed in China

102019

012020

THIS BOOK CONTAINS
RECYCLED MATERIALS

Photo Credits: Alamy, Animals Animals, Getty Images, iStock, Minden Pictures, Shutterstock

Production Contributors: Teddy Borth, Jennie Forsberg, Grace Hansen
Design Contributors: Dorothy Toth, Pakou Moua

Library of Congress Control Number: 2019941218
Publisher's Cataloging-in-Publication Data

Names: Hansen, Grace, author.

Title: Ermine / by Grace Hansen

Description: Minneapolis, Minnesota : Abdo Kids, 2020 | Series: Arctic animals | Includes online
   resources and index.

Identifiers: ISBN 9781532188862 (lib. bdg.) | ISBN 9781532189357 (ebook) | ISBN 9781098200336
   (Read-to-Me ebook)

Subjects: LCSH: Ermine--Juvenile literature. | Stoat--Juvenile literature. | Weasels--Juvenile literature. |
   Zoology--Arctic regions--Juvenile literature. | Arctic--Juvenile literature.

Classification: DDC 599.74447--dc23

Table of Contents

The Arctic

The Arctic is the northernmost part of Earth. It is made up of land, the Arctic Ocean, and the sea ice that floats on it. The weather there is freezing cold. Any animal that lives in the Arctic is tough!

Ermine

Ermines belong to the weasel family. Many of them call the Arctic home. In the Arctic, they live on **tundra** and rock and ice.

Ermines are small .

They grow to be only 5 to 12

inches (13 to 30 cm) long.

Their tails are no more than

5 inches (13 cm) long.

In wintertime, their coats
are thick and white. The tips
of their little tails are black.
Their coats keep them warm.

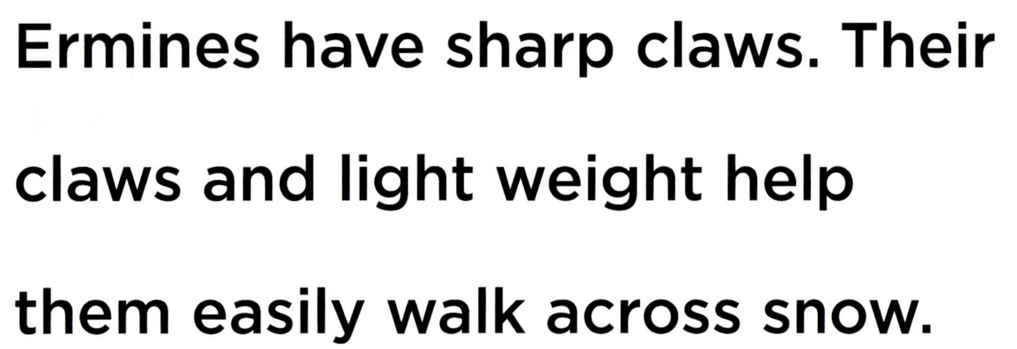

Ermines have sharp claws. Their claws and light weight help them easily walk across snow.

Ermines sleep during the day. They do most of their hunting at night. They like to eat small **mammals**, like mice. They also like to eat birds and eggs.

In the Arctic, ermines live in underground burrows. They also live in the **crevices** of rocks.

Baby Ermine

An ermine's home is called a
den. Females have their babies
in dens. A **litter** is often made
up of 3 to 18 **kits**.

18

19

The **kits** are blind and helpless at birth. They drink their mother's milk until they are 12 weeks old. They stay with their mother for about a year. During this time, they learn to survive.

More Facts

- Ermines that live in milder temperatures have fur that turns partially white.

- Ermines are quiet, sleek, and quick. These things make them great hunters.

- Because of its shape, an ermine can easily fit in the burrows of its prey.

Glossary

crevice – a narrow opening.

kit – the young of a weasel and some other furbearing mammals.

litter – a group of young animals born to one mother at one time.

mammal – a warm-blooded animal with fur or hair on its skin and a skeleton inside its body.

sea ice – frozen ocean water that is typically covered with snow.

tundra – one of the huge plains in the arctic regions of North America, Europe, and Asia. Trees do not grow on the tundra.

Index

Abdo Kids ONLINE

FREE! ONLINE MULTIMEDIA RESOURCES

Visit **abdokids.com** to access crafts, games, videos, and more!